YOUR KNOWLEDGE HAS VALUE

- We will publish your bachelor's and master's thesis, essays and papers

- Your own eBook and book - sold worldwide in all relevant shops

- Earn money with each sale

Upload your text at www.GRIN.com and publish for free

Bibliographic information published by the German National Library:

The German National Library lists this publication in the National Bibliography; detailed bibliographic data are available on the Internet at http://dnb.dnb.de .

Imprint:

Copyright © 2017 GRIN Verlag, Open Publishing GmbH
Print and binding: Books on Demand GmbH, Norderstedt Germany
ISBN: 9783668544345

This book at GRIN:

http://www.grin.com/en/e-book/376418/mutation-induced-in-gladiolas-through-physical-and-chemical-mutagens

Sudha Patil

Mutation Induced in Gladiolas through Physical and Chemical Mutagens

GRIN Publishing

Mutation induced in gladiolus (*Gladiolus hybridus*) cv. American Beauty through physical and chemical mutagens

Sudha Patil

Assistant Professor
Department of Floriculture and Landscape Architecture
ASPEE College of Horticulture and Forestry
Navsari Agricultural University, Navsari - 396 450 (Gujarat)

Abstract

The present investigation on "Mutation induced in gladiolus (*Gladiolus hybridus*) cv. American Beauty through physical and chemical mutagens" was carried out at Navsari Agricultural University, Navsari (Gujarat). Corms were treated with different doses of gamma rays (5, 6, 7, 8, 9 and 10 kR), EMS (0.5%, 1.0% and 1.5%) and DES (0.5%, 1.0% and 1.5%). Different vegetative, floral and yield characters were taken under study. Among all mutagenic treatments, early sprouting (10.40 days) and flowering (72.73 days) with maximum number of sprouts (2.27), plant height (58.86 cm), number of leaves (12.87), leaf area (33.90 cm^2), number of florets per spike (11.60), spike length (77.46 cm), floret size (11.04 cm), blooming period (11.27 days), number of spikes per plant (2.13) and number of corms (1.53) was recorded in treatment T_7 *i.e.* EMS @ 0.5%. Higher doses of all mutagens adversely affected growth, flowering and yield parameters of gladiolus.

Key words: Gamma rays, EMS, DES, gladiolus, mutation

Introduction

Gladiolus is known worldwide for its attractive spikes and has gained much importance as 'Queen of bulbous flowers' which has ever increasing demand in the flower market. Among the commercial flowers, gladiolus is one of the most important flower in India because of its majestic spikes containing attractive, elegant and delicate florets of various shades, sequential opening of flowers for a longer duration and good keeping quality of cut flowers (Singh, 2006). Gladiolus needs attention towards genetic improvement as the demand of flowers is increasing continuously. Since garden varieties of today come from diverse genetic parentage that are heteroploids ranging from 2n=30 to 180 and hypo aneauploids so the reproduction by seeds in this case has no meaning to maintain the varietal identity, but for evolution of new forms. Due to its heterozygosity in genetic constitutions, this makes it promising test material for induction of mutagens where only one or a few characters are to be improved upon without changing the entire genotype which offers promising possibilities. Gladiolus is grown vegetatively to perfection and so mutation breeding offers great potentialities as the mutated part can be conveniently perpetuated by vegetative means resulting in the development of new forms. Hence, the present investigation was conducted and the emphasis was laid on finding out desirable variations caused by physical mutagens (gamma radiations) and chemical mutagens (EMS & DES) in gladiolus cv. American Beauty.

Materials and Methods

The present experiment was carried out at the Floriculture Research Farm, Department of Floriculture and Landscaping, ASPEE College of Horticulture and Forestry, Navsari Agricultural University, Navsari, Gujarat, during winter season of year 2015-16. Total 13 treatments of gamma radiation, Ethyl Methan Sulphonate (EMS) and Diethyl Sulphonate (DES) were taken *viz.,* 5 kR gamma rays (T_1), 6 kR gamma rays (T_2), 7 kR gamma rays (T_3), 8 kR gamma rays (T_4), 9 kR gamma rays (T_5), 10 kR gamma rays (T_6), 0.5% EMS (T_7), 0.5% EMS (T_7), 1.0% EMS (T_8), 1.5% EMS (T_9), 0.5% DES (T_{10}), 1.0% DES (T_{11}), 1.5% DES (T_{12}) and control (T_{13}). The gladiolus corms (4 to 5 cm diameter) of cv. American Beauty were subjected to Co-60 gamma radiations at Bhabha Atomic Research Station, Trombay in October, 2015. The corms were treated with gamma radiations and planted within 24 hours of treatment. In case of chemical mutagens, corms were immersed in chemical solutions of three concentrations for 3 hours. After treatment, these corms were again immersed in STS solution for 15 minutes and then washed in running tap water for 20 minutes. Then the corms were planted on the raised beds in Randomized Block Design along with control (without treatment) and replicated thrice. Observations were recorded on all vegetative, flowering and yield parameters. Moreover pollen fertility, vegetative and flowering abnormalities were also studied. The observed data were statistically analysed by control versus rest method and the significance at 5% was observed (Panse and Sukhatme, 1967).

Results and Discussion

The data on different vegetative characters showed significant difference except plant height (Table 1). The data revealed that minimum days required for sprouting (10.40) with maximum number of sprouts per corm (2.27) was observed in treatment of 0.5% EMS which was at par with 5 kR gamma radiations (10.67 days and 2.13, respectively) as compared to rest and control. Maximum survival and sprouting per cent *i.e.* 100 was found when the corms were not treated. It was clearly seen that the days to sprouting were increasing while number of sprouts, sprouting per cent and survival per cent were found to be decreasing with the increasing concentration of gamma rays, EMS and DES. The decrease in sprouting at higher doses of the mutagens may be attributed to disturbances at cellular level (caused either at physiological (or) physical level) or because of hormonal activity. Similar results were also observed in petunia when treated with higher doses of EMS (Berenschot *et al.,* 2008). The uncertainity was observed because the alkylating agents are highly unstable entities. Moreover, the reduction in survival per cent is because of the effect of mutagen on meristematic cell or might be due to acute chromosomal damage, delay in onset of mitosis, chromosomal aberration induce enzyme activity such as catalyse and lipase production (Tiwari *et al.,* 2010). These results are in parallel line of work reported by Misra (1977) who also stated that higher doses of gamma radiations adversely affect both sprouting and survival of the plants in gladiolus cv. Oscar and Misra and Bajpai (1983) also reported similar results in gladiolus cv. Scarlet Double.

Looking at the growth of plant after the treatment of corms, plant height, number of leaves and leaf area, it was found that with the increasing dose of all mutagens, plant growth affected adversely. The maximum height was found in treatment of 0.5% EMS (58.86 cm) though there was no significant difference was observed in comparison to control. In case of leaves produced by plant and leaf area, significantly maximum values *i.e.* 12.87 and 33.90 cm^2, respectively were recorded in corms treated with EMS @ 0.5%. There was change in trend of leaf area where control was better over rest of the mutagenic treatments. This change may be because of the reason that alkylating agents are unstable. The increase in the growth at lower doses may because certain chemical mutagens produce single base substitutions with different mutation spectra because of which broad variation occur (Abdullah *et al.*, 2009). The plant growth reduces at higher doses of mutagens which might be because of destruction of growth inhibitors, drop of auxin level or inhibition of synthesis and decline of assimilation mechanism. Number of leaves and leaf area started to decreased as the dose of gamma rays increased irradiation produced minimum number of leaves per plant and leaf area. Such results have also been reported by Misra (1998) and Misra and Mahesh (1993) which may be due to activation of physiological substances present in corms at lower doses, while higher doses retard cell division by arresting mitotic cell division and causing ill effects on auxins.

The mean values of all the floral characters are presented in Table 2. Looking at the flowering characters, minimum days to first floret open (72.73 days) was observed in corms treated with EMS at 0.5% (T_7) which was at par with T_1, T_2, T_8 & T_9 whereas it was delayed with increase in doses and maximum number of days (92.93 days) was taken by corms which were treated with the dose of 10 kR gamma rays. Similar type of stimulatory effect was observed earlier by Misra and Bajpai (1983) in nine varieties of gladiolus and by Dhaduk (1992) with 5 kR in four varieties of gladiolus. Raghava *et al.* (1988) and Negi *et al.* (1983) noted that the flowering was delayed significantly at 5 kR treatment in various varieties of gladiolus due to higher chromosomal aberrations or synthesis of some toxic substances by mutagens that adversely affect the cell division. Similar results of delaying in flowering due to chemical mutagens were noted by Kapadiya (2014) in chrysanthemum cvs. Maghi and Jaya.

The results were found non significant with respect to number of florets per spike, length of spike, floret diameter and blooming period in control vs. rest but among all mutagenic treatments, significantly maximum number of florets per spike (11.60), length of spike (77.46 cm), floret diameter (11.04 cm) and blooming period (11.27 days) was recorded in corms treated with 0.5% EMS (T_7). The radiations reduced floret number, spike length, floret size and blooming period and affected adversely which may be because of auxin destruction, irregular auxin synthesis, failure of assimilation, mechanisms or inhibition of mitotic and chromosomal changes or damage with association of secondary physiological damage, which support the present findings (Banerji *et al.*, 1996). Dobanda (2004) also stated that higher doses of gamma rays showed deterioration in number of flowers opened simultaneously, because the length of phenological phases of gladiolus increased, which was directly proportional to the applied radiation doses. Among all treatments, the highest number of spikes (2.13) was observed in same treatment as compared to rest of the treatments and control. The number of spikes was decreasing with the increasing doses of mutagens which were supported by Kapadiya (2014) who found similar results when plants of chrysanthemum were treated with EMS and DES. High amount of variation is may be because of the chemical mutagens which produce single base substitutions with different mutation spectra due to which broad variation occur in yield parameters as compared to control (Khan *et al.*, 2009). Moreover, the alkylating agents are highly unstable which are responsible for sudden changes. In case of gamma rays treatments, similar results were reported by Sudha Patil (2009) with 6 kR and 7 kR doses which reduced yield of spikes drastically in cv. American Beauty while no flowering or few spikes were obtained in Nova Lux and Eurovision. Moreover, Raghava *et al.* (1981) found no flowering at 10 kR and 15 kR doses of gamma radiations in cvs. Little Gaint, Mansoer and Wild Rose which might be because of changes in plant metabolic activities and negative response of plant hormones to irradiations (Misra *et al.*, 2006).

Pollen fertility was studied in all treatments including control and it was found that the significantly maximum pollen fertility (90.16%) was recorded in control while it was drastically reduced in all mutagenic treatments and minimum fertility (26.51%) was obtained in application of 10 kR gamma rays. Among all mutagenic treatments, significantly maximum corms were produced by corms treated with EMS @ 0.5% while it was non significant when compared to control. Moreover, mutagenic treatments had adverse effect and produce vegetative and floral abnormality up to some extent. Maximum vegetative and floral abnormalities (1.39 and 1.34 %, respectively) were found in DES @ 1.5%. Such changes may occur due to physiological disturbances that may be because of reshuffling of histogen layers (Banerji and Datta, 2002).

References

Abdullah TE, Endan J, Mohd Naz B. Changes in flower development, chlorophyll mutation and alteration in plant morphology of *Curcuma alismatifolia* by gamma irradiation. *American J. Applied Sci.* 2009; 6 (7): 1436-1439.

Banerji BK, Datta SK. Induction and analysis of somatic mutation in ormanemtal bulbous plants. *J. Ornam. Horti.* (*New Series*). 2002; 5(1): 7-11.

Banerji BK, Dwivedi AK, Datta SK. Gamma irradiation studies on gladiolus. *J. Nuclear Agril. Biol.* 1996; 25 (2): 63-67.

Berenschot AM, Zucchi MI, Neto AT, Vera Q. Mutagenensis in *Petunia* × *hybrid* Vilm. And isolation of a novel morphological mutant. *Braz. J. Plant Physiol.* 2008; 2(2): 1590-1610.

Dhaduk BK. Induction of mutation in gar-den gladiolus by gamma rays. Ph. D. Thesis submitted to Post-Graduate School, Indian Agricultural Research Institute, New Delhi. 1992.

Dobanda E. Evaluation of variability induced by gamma radiation on quantitative and qualitative traits in gladiolus. *Cercetari de Genetica Vegetala si Animala.* 2004. **8**: 149-156.

Kapadiya DB. Induction of variability through mutagenesis in chrysanthemum (*Chrysanthemum morifolium* Ramat) varieties Jaya and Maghi through mutagenesis. *M.Sc. Thesis* submitted to Navsari Agricultural University, Navsari, Gujarat, India. 2014.

Khan S, Qurainy FA, Firoz Anwar. Sodium Azide: a chemical mutagen for exhibited a decline. *Int. J. Sci. Tech.* 2009; 4: 1-21.

Misra RL. Induction of mutations in gladiolus. *Plant Sci.* 1977; 9: 99-100.

Misra RL. Radiation induced variability in gladioli. *Indian J. Gen. and Plant Breeding.* 1998; 58(2): 237-239.

Misra RL, Mahesh KS. Studies on mutation induction in gladioli through Co^{60} gamma rays. *J. Ornam. Horti.* 1993; 1(2): 42-46.

Misra RL, Bajpai PN. Effect of mutagens on shooting, leaf number, heading, plant height and spike length in gladioli. *Indian J. Horti.* 1983; 7(1): 107-110.

Misra RL, Kumar N, Dhiman MR. Breeding perspective of ornamental bulbous crops. *In: Book of Abstracts of National Symposium on Ornamental Bulbous Crops*, SVPUA & T, Meerut, U.P., India, 5-6 December, 2006. pp. 1-6.

Negi SS, Raghava SPS, Sharma TVRS. Induction of mutations in gladiolus by gamma rays, Abstract of contributed paper XV International Congress of Genetics, New Delhi. 1983;

Dec. 12-27 Part II. Session C-IVD to VIID : 480.

Panse VJ, Sukhatme PV. Statistical Methods for Agricultural Workers, Indian Council of Agricultural Research, New Delhi. 1967.

Raghava SPS, Negi SS, Sharma TVRS. New cultivars of gladiolus. *Indian J. Hort.* 1981; 26(3): 2-3.

Raghava SPS, Negi SS, Sharma TVRS, Balakrishnan KA. Gamma ray in-duced mutants in gladiolus. *J. Nuclear Agril. Biol.* 1988; 17(1): 5-10.

Singh AK. Flower crops: Cultivation and Management. *New Delhi Publishing Agency,* 2006; 147.

Sudha Patil. Gamma rays induced mutations in commercial varieties of gladiolus (*Gladiolus hybrida* L.). *Ph.D. Thesis* submitted to Navsari Agricultural University, Navsari, Gujarat, India. 2009.

Tiwari AK, Srivastava RM, Kumar V, Yadav LB, Misra SK. Gamma rays induced morphological changes in gladiolus. *Prog. Agric.* 2010; 10 (special issue): 75-82.

Table: 1. Effect of different mutagenic treatments on gladiolus var. American Beauty

Treat-ments	Days to sprouting	Sprouts per corm	Sprouting per cent	Survival per cent	Plant height (cm)	Leaves per plant	Leaf area (cm^2)
T_1	10.67	2.13	9.98 (99.17)	9.89 (97.50)	51.77	11.80	29.57
T_2	13.00	1.80	9.93 (98.33)	9.73 (94.17)	49.43	11.20	24.43
T_3	16.10	1.80	9.89 (97.50)	9.46 (89.17)	43.37	11.13	26.47
T_4	21.40	1.53	9.89 (97.50)	9.38 (87.50)	42.67	10.93	25.03
T_5	23.67	1.47	9.81 (95.83)	9.15 (83.33)	39.99	10.80	21.20
T_6	24.73	1.33	9.64 (92.50)	9.11 (82.50)	34.04	9.47	19.49
T_7	10.40	2.27	10.02 (100)	9.93 (98.33)	58.86	12.87	33.90
T_8	12.20	1.87	9.98 (99.17)	9.89 (97.50)	53.96	11.73	32.66
T_9	12.27	1.80	9.85 (96.67)	9.81 (95.83)	46.72	11.27	30.56
T_{10}	13.47	1.80	9.72 (94.17)	9.68 (93.33)	47.21	10.53	28.96
T_{11}	14.13	1.47	9.68 (93.33)	9.60 (91.67)	41.19	9.60	24.08
T_{12}	14.93	1.67	9.55 (90.83)	9.29 (85.83)	36.54	8.40	21.54
CD at 5%	1.26	0.42	0.21	0.16	9.14	0.94	6.35
T_{13} - Control	10.93	2.07	10.02 (100)	10.02 (100)	48.52	10.07	32.74
CD at 5%	0.93	0.31	0.16	0.12	NS	0.70	4.67
CV %	4.94	14.10	1.30	0.97	11.93	5.24	14.05

Table 2: Effect of different mutagenic treatments on gladiolus var. American Beauty

Treat-ments	Days to first floret opening	Florets per spikes	Spikes length (cm)	Floret diameter (cm)	Blooming period (day)	Spikes per plant	Pollen fertility (%)	Corms per plant	Vegeta-tive abnormality (%)	Floral abnor-mality (%)
T_1	75.60	9.93	57.92	8.60	9.40	1.80	58.59	1.53	0.70	0.70
T_2	77.00	9.33	55.11	8.50	9.20	1.67	56.77	1.40	0.70	0.70
T_3	79.27	8.93	47.77	7.41	8.60	1.20	47.77	1.67	0.70	0.70
T_4	84.80	8.87	46.97	7.23	8.47	1.00	46.97	1.33	0.70	0.70
T_5	86.47	7.67	37.21	6.36	8.07	0.60	41.55	1.27	0.70	0.70
T_6	92.93	6.67	26.51	6.06	7.87	0.60	26.51	1.20	1.04	1.05
T_7	72.73	11.60	77.46	11.04	11.27	2.13	77.46	1.93	0.70	0.70
T_8	76.53	11.47	70.30	10.55	10.13	1.73	70.30	1.53	0.70	0.70
T_9	76.47	11.27	68.73	9.49	9.53	1.60	68.73	1.27	0.70	0.70
T_{10}	82.40	9.73	58.06	8.85	9.47	1.47	58.06	1.67	0.70	0.70
T_{11}	84.07	8.80	49.62	7.85	8.60	1.33	52.96	1.47	0.70	0.87
T_{12}	86.90	7.20	44.85	7.57	8.00	1.20	46.19	1.20	1.39	1.34
CD at 5%	7.02	1.09	9.53	1.71	1.77	0.41	10.15	0.38	0.36	0.21
T_{13}-Control	83.73	9.87	52.16	9.14	10.07	1.87	90.16	1.40	0.70	0.70
CD at 5%	5.16	NS	NS	NS	NS	0.30	7.47	NS	NS	NS
CV %	5.13	6.95	10.68	12.19	11.54	17.51	10.61	15.76	27.63	16.18

YOUR KNOWLEDGE HAS VALUE

- We will publish your bachelor's and
  master's thesis, essays and papers

- Your own eBook and book -
  sold worldwide in all relevant shops

- Earn money with each sale

Upload your text at www.GRIN.com
and publish for free